James Cook

THE OPENING OF THE PACIFIC

Basil Greenhill, CMG
DIRECTOR,
NATIONAL MARITIME MUSEUM

Y0-DBX-030

James Cook's early life

Youth

Born at Marton in Yorkshire in 1729, the son of a farm labourer turned overseer, at first a labourer himself, he learned to read and write at the village school at Great Ayton and became a grocer's assistant at the seaside village of Staithes.

He soon went to sea in a collier to improve his lot by becoming a master and, in due course, part owner of a ship.

Joins the Navy, 1755

In 1755, when in a position to take command of a merchant ship he joined the Royal Navy, being promoted soon to a senior warrant officer.

Becomes a proficient hydrographer, 1758–59

His commanding officer, Captain Palliser, later Sir Hugh and Comptroller of the Navy and a very influential friend, recognised his great ability. In 1758 and 59 he learned survey work from Army officers in Nova Scotia during the Seven Years War, 1756-63, against France. He helped in the charting of the St. Lawrence for the capture of Quebec by General Wolfe in 1759.

Charts Newfoundland, 1763–67

With the ending of the war, Cook, now in command of the schooner *Grenville*, spent four more years charting the coasts of Newfoundland, whose cod fisheries were of great economic importance.

Chosen to explore the Pacific, 1768

In 1761 and again in 1769 the planet Venus passed between the Earth and the Sun. By observing from different parts of the Earth the passage of the planet Venus across the disk of the sun astronomers hoped to calculate the distance of the earth from the Sun and thus have a means to measure the size of the solar system.

James Cook was selected in 1768 as a Lieutenant to command a scientific expedition to the newly-discovered island of Tahiti where a transit of Venus observation was to be made in 1769.

He had secret orders directing him afterwards to seek the Great Southern Continent which Captain Wallis thought he had seen to the south of Tahiti in 1767.

Captain Hugh Palliser, later Admiral of the White (1723–1796)
Oil painting attributed to George Dance (1741–1825) after Sir Nathaniel Dance (1735–1811)
Impressed with Cook's ability as a surveyor, Palliser was largely responsible for his appointment to command the expedition to the Pacific in 1769

The opening of the Pacific

For James Cook, 'the most able and enlightened navigator that England ever produced', as a contemporary described him, the times were ripe, as they have been for few men in history. He came up from the lower deck of the Navy little known and with few connections in a society in which connections were very important. In ten years he had explored more of the earth's surface than any other man in history. As Professor J. C. Beaglehole, who has done more than anyone else to reveal the extent and nature of Cook's achievement, has said, 'his competence changed the face of the world'. He wrote little of his private thoughts. Almost no private papers survive and very little is known of his personal life. But there is enough to relate him to his times. The Cook Gallery in the National Maritime Museum seeks to convey this relation to the visitor.

For centuries many people in Europe had believed that there must be a great southern continent stretching far north into the temperate zone of the South Pacific. Two hundred years ago there were still no certainties about what was there. There were the technical difficulties of sailing a ship across the ocean from east to west in high latitudes in the face of the prevailing wind system and of fixing the position accurately of a ship and of lands and islands discovered in the ocean. Besides these very important obstacles to Pacific exploration the social and intellectual climates of European countries before the second half of the eighteenth century did not tend to produce the conditions and the men needed to operate a sailing vessel for long periods at extreme range from its base, and the vast size of the Pacific required exploration for years on end without any possibility of support.

When Captain Samuel Wallis returned to England in 1768, the latest of the long series of unsuccessful explorers of the South Pacific, his men reported that they had sighted near Tahiti (which he had discovered and been able to fix accurately by the lunar distance method of determining longitude recently perfected at the Royal Observatory at Greenwich, the old buildings of which are now part of the National Maritime Museum), 'the long wished for southern continent, which has often been talked of, but never before seen by any Europeans'. Britain could not afford to have another European power firmly established in this southern continent, if it existed. She was at the beginning of the commercial and industrial revolution which followed upon the technological discoveries of the early eighteenth century. Her great rival, France, had lost her empire in North America and India to Britain and was looking for compensatory discoveries elsewhere. Whoever laid effective claim to a new productive continent would be a world leader for generations. Three months before Wallis' return the Royal Society petitioned King George III for support in observing from, (among other places) the South Seas, the Transit of Venus between the earth and the sun, due to occur in 1769. From these observations it was thought it would be possible to calculate the distance of the earth from the sun, an important stage in determining the size of the Universe.

In 1768 James Cook, with seventeen years as a merchant seaman behind him and now a Master in the Royal Navy in command of the schooner *Grenville*, had completed nearly six years of survey work on the coast of Newfoundland – important and conspicuous work because of the value of the cod fishery, much in dispute with France. Before his Newfoundland surveys Cook had distinguished himself on survey work in the St. Lawrence. Although still not commissioned he was now recognised as the best of the few hydrographic surveyors in the Royal Navy. Admiral Hugh Palliser, one of Cook's first captains during his early years in the Navy, and Governor of Newfoundland while Cook was there, was now Comptroller of the Navy Board; he and others in the Admiralty with whom the decisions lay for appointing a Commander for the Pacific expedition appreciated Cook's peculiar talents for the post. They appear to have been determined against the Royal Society's civilian candidate for command of the South Seas expedition, Alexander Dalrymple, who, besides being an experienced hydrographer of the East India Company, was also the leading contemporary exponent of the theory that a great continent existed in the southern hemisphere. Cook's record as a seaman, astronomer and hydrographer made him the best candidate, in the Admiralty's view, for command of a small ship engaged in unspectacular work which involved not only accurate and painstaking astronomical observation but also, if the continent was found (and he was secretly instructed to search for it), careful surveying and charting. They chose accordingly.

James Cook made three great voyages. He started on his first with certain great advantages. The vessel purchased for the expedition in the absence of an available Royal Naval ship was a merchantman peculiarly suited to operate at extreme range for long periods because she was very roomy and at the same time was small enough, in that age of the perfection of the carpenter's and blacksmith's

4

technology, to be maintained almost indefinitely by her own crew without dockyard support. She was of large capacity for her size, yet shallow and able to take the ground in tidal water and remain more or less upright, an important factor in exploring unknown shallow coasts. She could be put deliberately aground in tidal water in order that her crew could do necessary maintenance work on her hull without a dry-dock. A larger vessel with sharper sections could not be laid ashore safely.

Two years before Cook's first voyage began the first edition of the *Nautical Almanac* had been published with tables which at long, long last enabled the mariner relatively easily, with accuracy to calculate his longitude, that is, his position on the earth's surface east or west of Greenwich to about thirty miles. Before this he could determine only his latitude, that is, his position north and south to this accuracy, by observation of the sun, using a quadrant. His longitude, east and west, he could only guess by estimating his distance and direction sailed. Some seamen were very skilful at this, but particularly because of unknown currents, on long voyages knowledge of the ship's position became very approximate. To find an island or a place on a coastline the navigator had to sail to its latitude and then sail down the latitude until he hit the place he was looking for – he hoped not too hard! He also hoped he was sailing in the right direction: i.e. not leaving his objective astern of him. Often the eventual landfall was violent and until far into the nineteenth century many vessels were lost by this kind of latitude navigation. Not only did the navigator not know where he was under such conditions, but, from the same causes his predecessors who had made discoveries had not been able to determine where they had been. Consequently the prudent navigator took no risks when sailing off coasts or among strange islands.

But Cook was able to take risks because, by using Hadley reflecting quadrants and a sextant (both recent inventions), a pocket watch and the *Nautical Almanac* he could sail straight for an intended landfall, once its position had been established, and then survey and chart coastlines more accurately than any explorer had ever done before.

On his second voyage, which was perhaps the greatest sailing ship voyage ever made, Cook had with him Larcum Kendall's copy of Harrison's perfected chronometer (both of which are still ticking away in the National Maritime Museum), which gave him positions to within about three miles; he was thus the first commander of a ship in history to know very closely where he was for most of his time at sea. He could return with relative ease to the island bases on which he depended. Moreover, these were almost the first British voyages of exploration to have professional astronomers among their people. Provided by the Royal Society and the Board of Longitude with the best instruments of their day these men had as their main function to settle precisely the latitude and longitude of places on shore.

On all three voyages the Admiralty was anxious to try, or to try again, possible methods of avoiding scurvy, and Cook was liberally supplied with supposed anti-scorbutics. He rigorously enforced their use, as he enforced strict cleanliness on his crews. The effect was no less than miraculous and Cook's expeditions were not weakened by disastrous losses of men from the deficiency diseases which had ruined so many previous expeditions.

From all three voyages Cook brought back a wealth of scientific data and specimens on a scale never before equalled. Not all of it has been fully worked over, even today. An enormous number of astronomical observations were made, recorded and published. A vast amount of survey work was done, to the degree that after the completion of the third voyage it could be said that the Pacific had been explored and, for the first time, accurately charted. Thousands of biological and botanical specimens were collected and thousands of drawings and descriptions prepared by the artists who accompanied each expedition. The artists also recorded

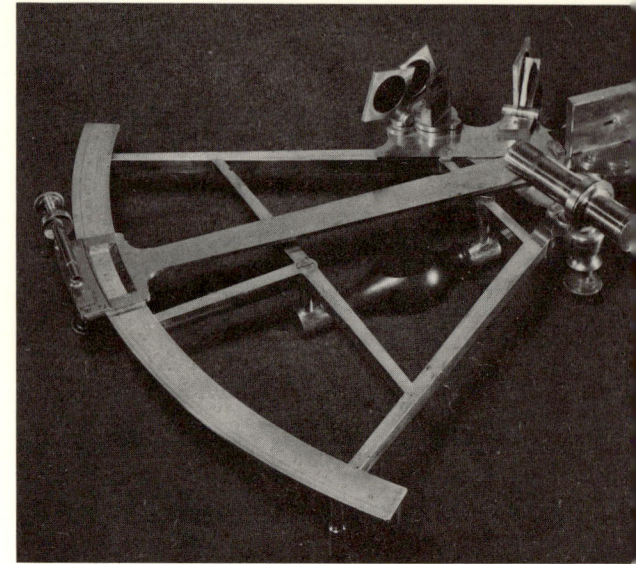

Sextant, about 1775
by Ramsden
Of the type used by Cook and his astronomers for lunar distance and other observations of the Sun, Moon and stars

Larcum Kendall's first marine chronometer (K1), 1769
Cook took this chronometer on his second and third voyages during which time it performed reliably under rigorous conditions

Dipping Needle, 1772
by Nairne
Said to have been taken on Cook's second voyage, to measure the vertical component of the Earth's magnetic field

the passing scene, acting somewhat as expedition photographers would do today. Under the combined demands of seamen and scientists for accuracy they drew and painted naturalistic pictures which started a trend, culminating in the realism of the nineteenth century. The artistic products of Cook's voyages, and those in particular of William Hodges, the artist of the second voyage, had widespread cultural and artistic influence.

On the first voyage Cook surveyed New Zealand, which had been discovered by Abel Tasman in 1642, and established that it was not part of a great southern continent. He also narrowed the possible limits of such a continent by systematic exploration of the east and west sides of the South Pacific. He discovered and surveyed the whole of the east coast of Australia, for only parts of the south, west and north, from Dutch exploration, were already known.

On the second voyage Cook finally obliterated the vision of a southern continent, sailing backwards and forwards over a great part of the area it was supposed to occupy. He showed that Antarctica (which he suspected to exist) must be a frigid outpost not normally habitable.

On the third voyage he discovered Hawaii, surveyed three thousand miles of the west coast of North America and thrust through the Bering Strait into the Arctic until stopped by the ice. He showed that no practicable passage, navigable by sailing vessels from the Pacific to the Atlantic existed round or through North America.

That he was born into a humble position in eighteenth century society – his father was a farm labourer who bettered himself and became an overseer and he himself was first a labourer and then a shop assistant before he went to sea – may have been one of James Cook's great sources of strength. Perhaps he lacked prejudices and patterns of thought sometimes instilled by the more traditional upbringing and the classical

education of the time. His immediate predecessors, Byron and Wallis, had set out, as he did on his first voyage, with orders to search for the fabled southern continent, but they had been forced north, away from the unexplored southern waters where it might lie. Cook's origins were not in the order-giving classes and this fact may in part have influenced the thoroughness with which he executed his instructions. Very practical attitudes of mind were necessary to use the new techniques against scurvy – readiness to experiment, determination to enforce unfamiliar routines and diets upon eighteenth century seamen possessed of all the extreme conservatism of the very simple men many of them were, and meticulous and unceasing attention to detail. It fell to the highly intelligent former merchant seaman whose naval career had been largely spent in acquiring and exercising the new and exacting skills of the surveyor to be the first man to apply the new techniques successfully.

Cook as a seaman had other special attributes born of his origins. Leaving a shore job to go to sea in the coasting trade was not at the time such an unusual way to escape the treadmill of relative poverty. The seaman could become mate, master, shareholder, small capitalist. But in the days of sailing ships there were, broadly speaking, two kinds of professional, the deepwater man, who disliked coasts and sought to avoid contact with them until he had to make his way into harbour, and the coasting seaman, intimately concerned with tides, shoals, narrow passages, baffling winds, pilotage. Unlike most of his contemporaries Cook was conditioned by his experience for both roles and without the techniques of the coasting seaman he might well have turned away from the very waters, off New Zealand, off Australia, among the Pacific Islands, off the West Coasts of Canada and the State of Alaska, where his positive achievement was greatest.

James Cook had a highly developed sense of responsibility towards his fellow men – the men under his command, those who had sent him out, those who would follow, as well as to the primitive societies which he discovered and upon which, as he foresaw, the impact of Europe was to prove disastrous. His very tenacity and singleness of purpose may have had something to do with the limited circumstances of his youth. A sensitive and sensible man, he lived, during his first great voyage, for years with men of the highest intellectual calibre, trained above their contemporaries in science and the humanities, Banks, Solander, Green the astronomer, in the enforced intimacy imposed by the very cramped quarters of a small eighteenth century vessel. The society of these young men from a different world undoubtedly sharpened Cook's latent intellectual abilities, broadened his sympathies. The Cook of the second voyage was the product partly of this educational process.

But this society, and that of William Wales, the astronomer of the second voyage, and of the Forsters, father and son, the scientists of the second voyage, whatever their merits and demerits, spared him the loneliness of command. Only on the third voyage was he in the position so familiar to his contemporaries in the Navy, that is without the easy day-to-day society of men of comparable intellectual calibre who were not subordinate to him in the naval hierarchy. Added to the strains of the preceding years this deficiency may have contributed to errors of judgment and to the final disaster of his death in a meaningless fracas on Hawaii.

Cook's voyages lie at a watershed in history. The first man to have the opportunity to do so many things, it was supremely fortunate that he had the ability and character to see and exploit the possibilities. Before him no comparable voyage had been made. After him, with the advancing industrial revolution, world voyaging in merchant ships soon became commonplace. Colonisation spread throughout Australasia and the South Seas. Inevitably at the end of the eighteenth century, sooner or later, the world would have begun the contracting process which has been accelerating

ever since. One of James Cook's greatest achievements was that, whatever may have followed, this acceleration was initiated so decisively and with such credit to all who were concerned with these three great voyages, in which man for the first time determined the shape and limits of the habitable earth.

John, 4th Earl of Sandwich (1718–1792)
Oil painting by Thomas Gainsborough (1727–1788)
To Cook he was both friend and patron, and though he made many enemies by confounding his private with his political life, he planned the grand strategy of Cook's explorations, recognised Cook's genius and furthered his endeavours unfailingly

First Voyage
1768-1771

Ship

HMS ENDEAVOUR
366 tons

Lieutenant James Cook
5 officers and 88 men

Joseph Banks, naturalist
Daniel Carl Solander, naturalist
Charles Green, astronomer
Sydney Parkinson, artist
Alexander Buchan, artist
Diedrich Herman Spöring, assistant naturalist

Total 100 men

Model of HMS ENDEAVOUR made in 1969 by craftsmen of the National Maritime Museum

She was commanded by Lieutenant James Cook on his first voyage to the Pacific and sailed from Deptford on 21 July 1768. A cat-built bark, she was of large capacity, but at the same time small enough to be maintained by her crew without dockyard support and therefore particularly suited for long range exploration

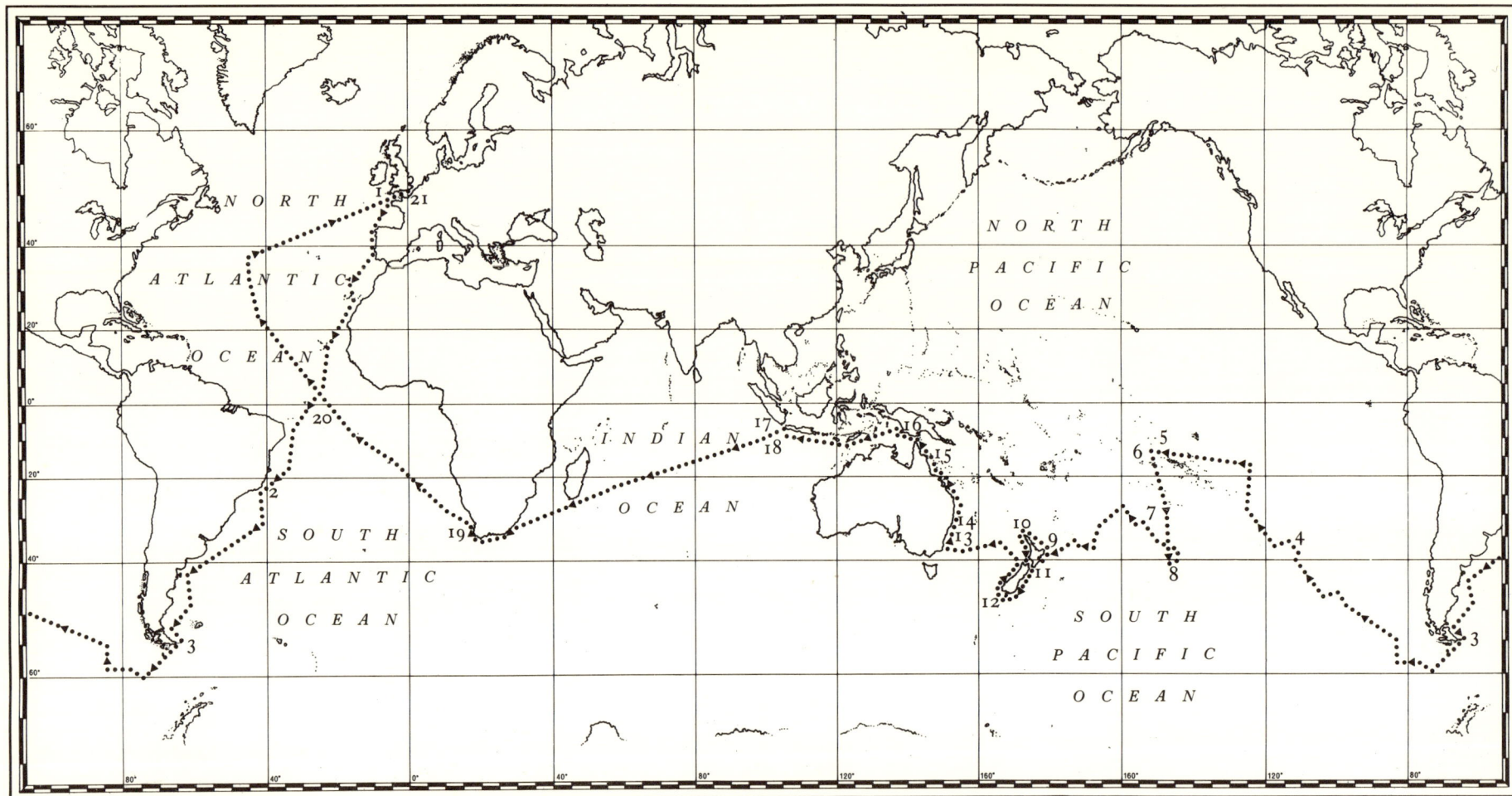

Sir Joseph Banks, BART (1743–1820) *left*
Detail from a mezzotint by J. R. Smith after
Benjamin West, 1788

Daniel Solander, naturalist (1733–1782) *centre*
from the London Magazine, 1772
*Banks invited him to join the first expedition as chief
naturalist*

Sydney Parkinson, artist (about 1745–1771) *right*
from 'A Journal of a Voyage to the South Seas . . .'
S. Parkinson, London, 1773
The most important artist on the first voyage

KEY	DATE	POSITION
	1768	
1	*August*	Sailed from Plymouth
2	*December*	Rio de Janeiro
	1769	
3	*January*	Cape Horn
4	*March*	South Pacific
5	*April*	Arrived Tahiti
6	*June–July*	Observed Transit of Venus at Tahiti and explored Society Islands
7	*August–October*	Searched westward for Great Southern Continent; results negative
8	*September*	Latitude 40°S (1st SEPT)
9	*October*	Arrived New Zealand, Young Nick's Head
10	*October–December*	Charting North Island, New Zealand
	1770	
11	*January*	Refit in New Zealand, Queen Charlotte Sound
12	*February–April*	Charting South Island, New Zealand
13	*April*	Arrived east coast of Australia
14	*May*	Botany Bay
15	*June*	Endeavour under repair at Endeavour River after stranding on reef
16	*September*	New Guinea
17	*October*	Batavia
18	*December*	Sailed for England
	1771	
19	*March*	Cape Town
20	*May*	South Atlantic
21	*July*	Arrived Downs

To observe the passage of the planet Venus over the disk of the Sun on the 3rd June, 1769. (Transit of Venus)

Then, to search for an imagined continent south of Tahiti and westward to New Zealand

Then to chart the coast of New Zealand and any discovered islands

Achievements

Observed the Transit of Venus

Searched for and proved there was no continent to the south and westward of Tahiti, north of latitude 40° South, as far as New Zealand

Discovered the east coast and charted all the coasts of New Zealand

Discovered and charted the east coast of Australia

On this first voyage, Cook was accompanied by the wealthy young Mr Joseph Banks aged 25, a Fellow and later President of the Royal Society, a keen natural historian and patron of science. Banks undertook the world voyage partly in place of making the fashionable Grand Tour of European capitals and partly to further scientific knowledge.

At his own expense Banks took with him naturalists and natural history artists to record in faithful and methodical detail the fauna, flora, sea-life, and peoples met with and their manner of living, both to provide a firmer basis for natural philosophy and to entertain his friends on his return.

Unfortunately, one artist, Buchan, died when Tahiti was reached in 1769, and the other, Parkinson, after leaving Batavia on the voyage home. It is, therefore, chiefly from published engravings of Parkinson's works made by other artists that western society got its first visual impression of the hitherto virtually unknown world of the South Pacific.

View of Matavai Bay from One Tree Hill, Tahiti, 1769
An engraving, after a drawing by S. Parkinson
The ENDEAVOUR *can be seen at anchor off Fort Venus*

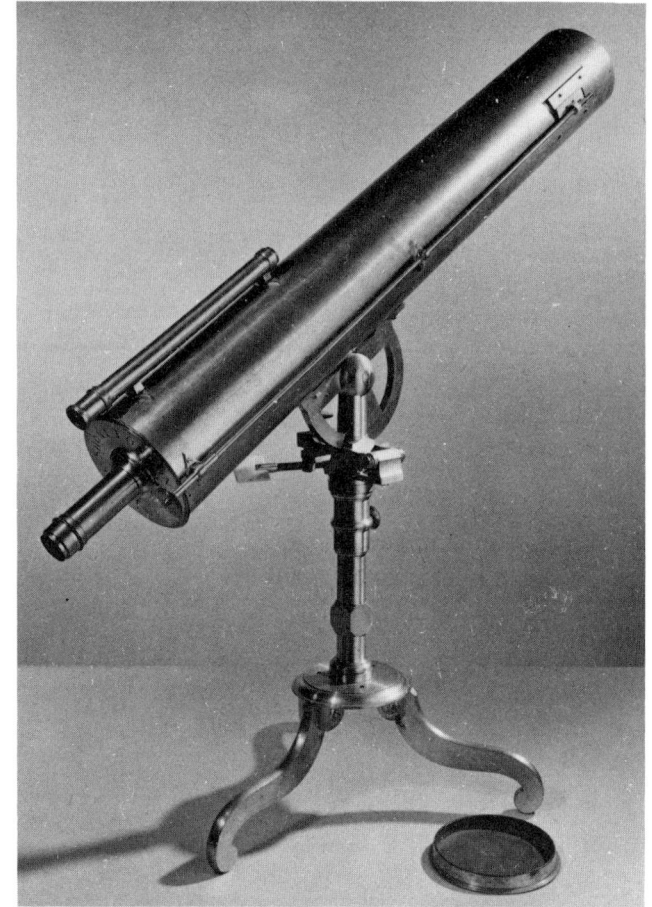

2ft focus Gregorian Reflecting Telescope, 1763
by James Short
Of the type used to observe the transit of Venus at Tahiti in 1769

Property of the Whipple Science Museum, Cambridge

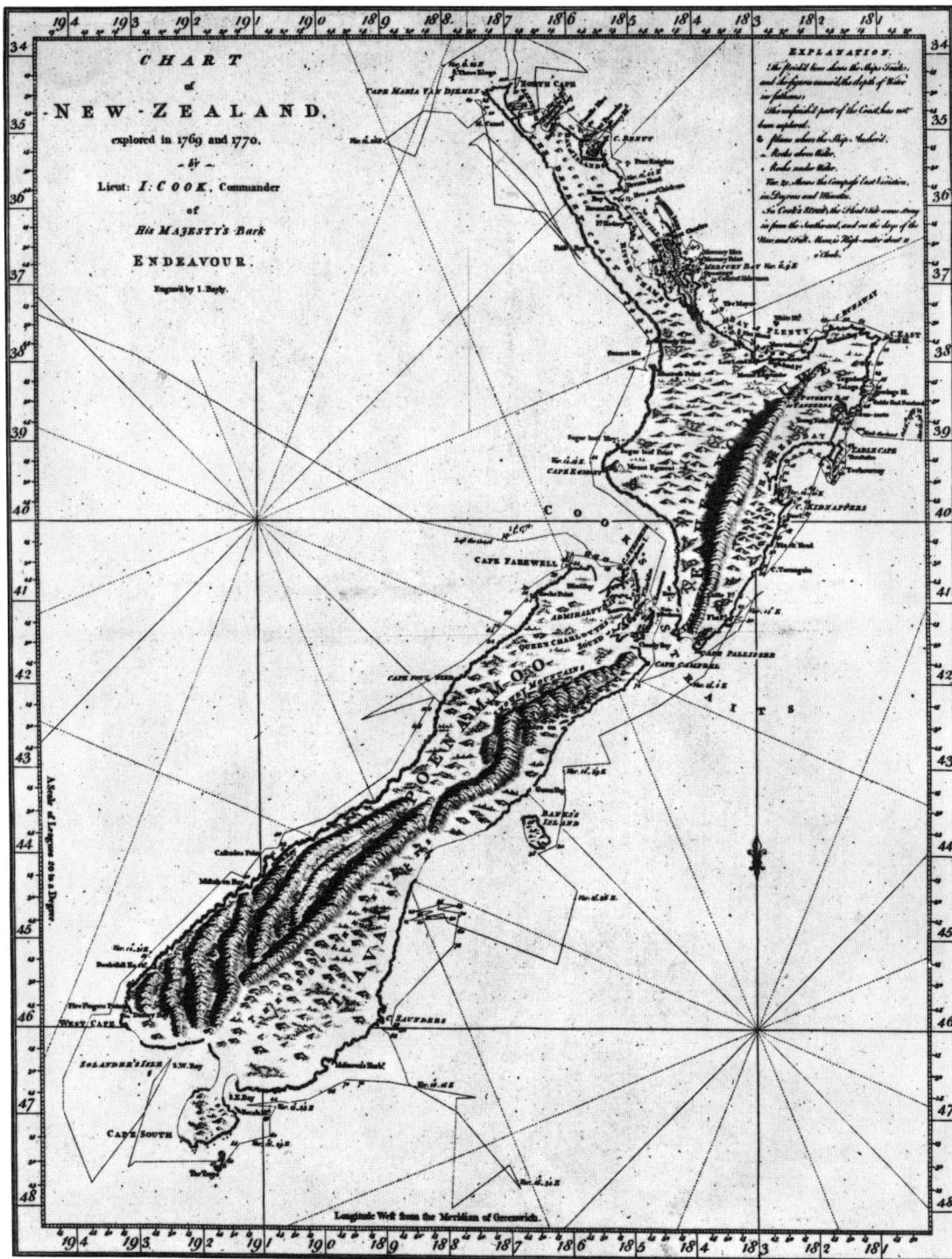

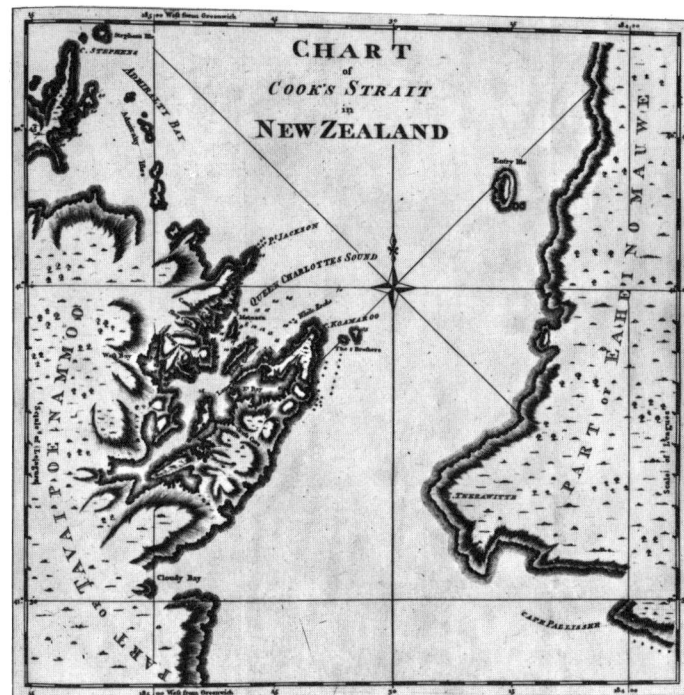

Cook's Strait, New Zealand, 1769
from Hawkesworth 'An Account of the Voyages . . .
in the Southern Hemisphere' London 1773
The chart shows Cook's anchorages in Queen Charlotte's Sound

New Zealand, 1769
Engraved chart by I. Bayly after J. Cook (published 1772)
The chart shows the track of Cook's first voyage clearly

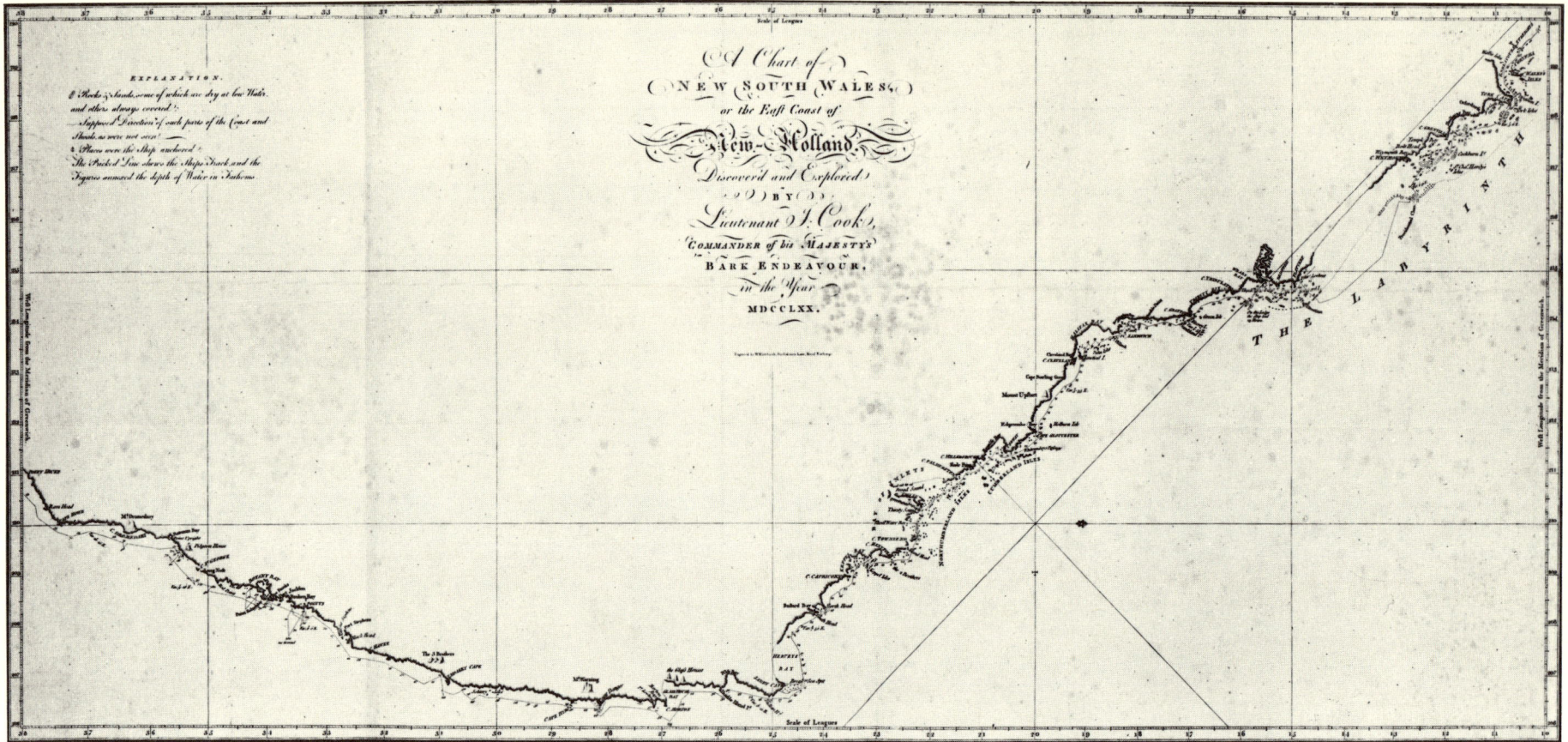

A Chart of NEW SOUTH WALES or the East Coast of New Holland, Discover'd and Explored by Lieutenant I. Cook, Commander of his Majesty's Bark Endeavour, in the Year MDCCLXX

THE LABYRINTH

East coast of Australia, 1770

Engraved chart by W. Whitchurch after J. Cook (published 1772)

Cook was the first European to explore the eastern coast of New Holland, as Australia was then known. He called it all New South Wales although the northern part of this coast is now Queensland

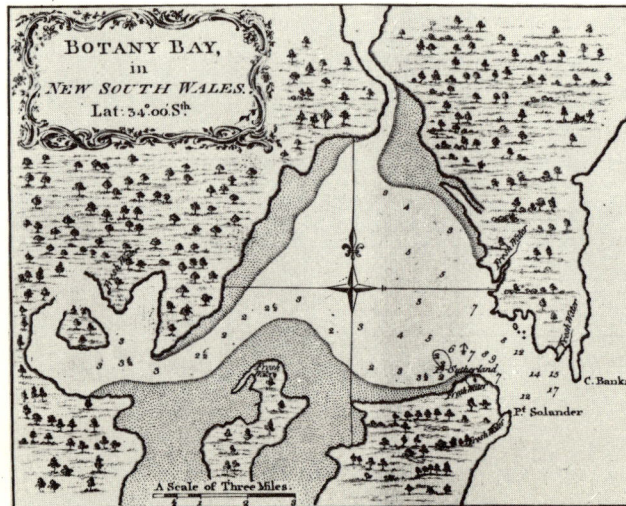

BOTANY BAY, in NEW SOUTH WALES. Lat: 34°.00'S'th

A Scale of Three Miles.

ENTRANCE of ENDEAVOUR RIVER, in NEW SOUTH WALES. Lat: 15° 26' S'th

A. The place where we landed our Stores.
B. .... Repair'd the Ship.

The figures denote the depth in fathoms at low Water.

Mangroves

A Scale of one Mile.

Botany Bay, 1770

Engraved by J. Gibson and T. Bowen

A view of the Endeavour River on the east coast of Australia, 1770
An engraving by W. Byrne
The ENDEAVOUR *is shown 'laid on shore, in order to*
repair the damage which she received on the rock' in June 1770

Endeavour River, 1770
Engraved by J. Gibson and T. Bowen

Kangaroo, Endeavour River, 23 June 1770
Engraving after a drawing by S. Parkinson

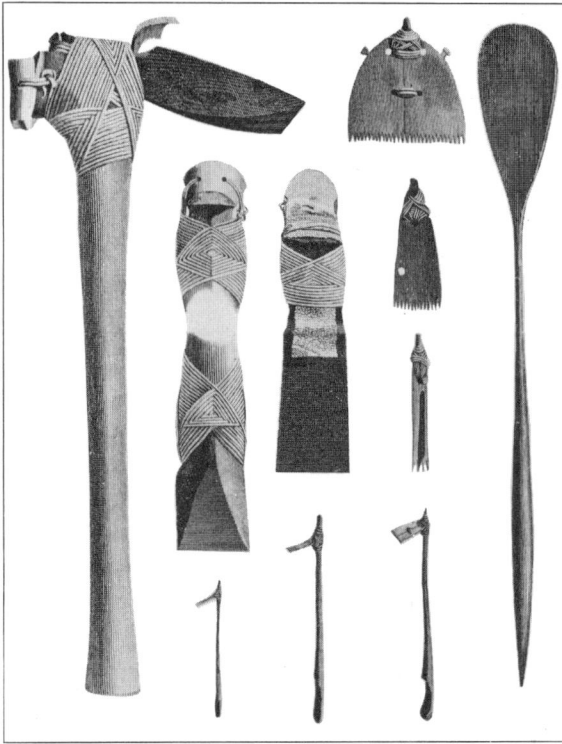

Stone tools and tatooing instruments, Tahiti, 1769
Engraved by Record

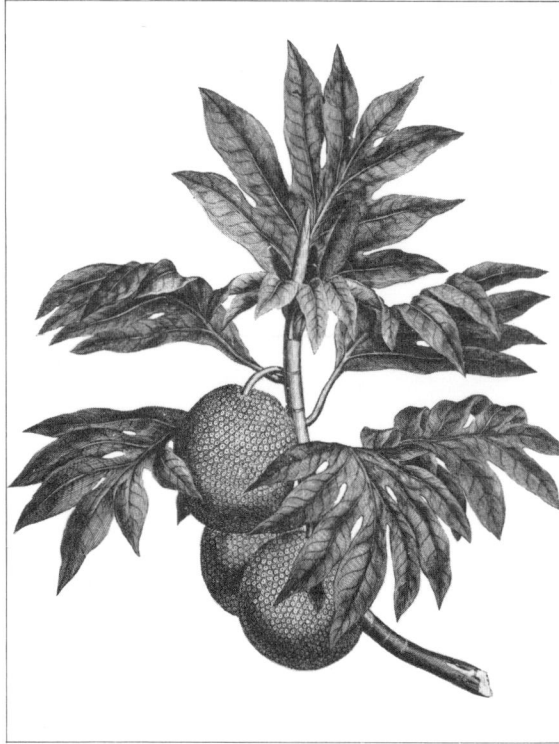

Bread fruit, Tahiti, 1769
Engraved by J. Miller

Head of Maori, New Zealand, 1769
An engraving, after a drawing by S. Parkinson
*This Maori chieftain is shown with 'a comb in his hair,
an ornament of green stone in his ear and another of
a fish's tooth round his neck'*

Second Voyage
1772-1775

Ships

HMS RESOLUTION
462 tons

Captain James Cook
8 officers and 102 men

William Wales, astronomer
Johann Forster, botanist
Georg Forster, botanist
William Hodges, artist
Two servants

Total 117 men

HMS ADVENTURE
366 tons

Captain Tobias Furneaux
5 officers and 75 men

William Bayly, astronomer
One servant

Total 83 men

Southern Hemisphere
Engraved by W. Whitchurch after George Forster, FRS
The chart shows the tracks of the RESOLUTION and ADVENTURE during Cook's second voyage, when he finally proved that the fabled Southern Continent did not exist

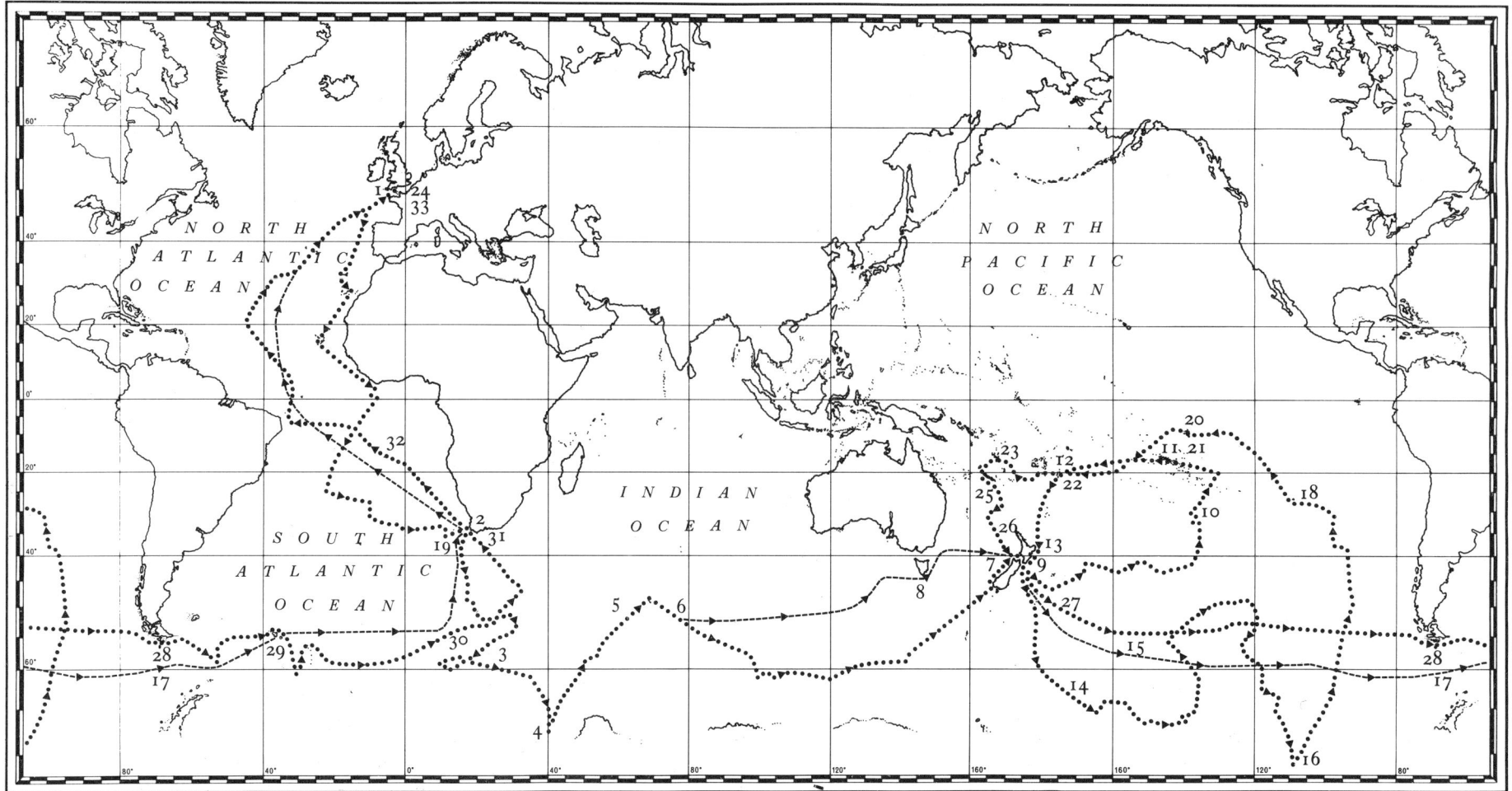

William Wales, FRS, astronomer (about 1734–1798)
Pastel by John Russell

In the possession of Lieutenant Colonel D. St. J. Edwards

William Hodges, artist (1744–1797)   *centre*
A drawing by George Dance (1741–1825)

Property of the Royal Academy of Arts

Omai, about 1773   *right*
Engraving by J. Caldwell after William Hodges
*Omai was a Society islander brought to England at his
own request by Captain Furneaux in the* ADVENTURE.
Cook returned Omai home on the third voyage

Voyage

KEY	DATE	POSITION
	1772	
1	*July*	Sailed from Plymouth
2	*October*	Cape Town
3	*November*	Sailed from Cape Town on Antarctic ice-edge search
	1773	
4	*January*	Latitude 67° 15′ S (18th JAN)
5	*February–*	South Indian Ocean
6	*March*	*Separated, Kerguelen Islands**
7	*March*	Arrived New Zealand
8		*Tasmania*
9	*May*	*Re-joined in Queen Charlotte's Sound*
10	*June–October*	First Tropical (inner) sweep
11	*August*	Tahiti
12	*October*	Tongan Islands
13		*Returned to New Zealand*
14	*November*	Second Antarctic ice-edge search
15	*December*	*Sailed from New Zealand because of insufficient stores. Searched South Pacific sector of Antarctic*
	1774	
16	*January*	Latitude 71° 10′ S (30th JAN)
17		*Cape Horn. Searched South Atlantic sector of Antarctic*
18	*March*	Easter Island. Second Tropical (outer) sweep
19		*Cape Town*
20	*April*	Marquesas Islands
21	*April*	Tahiti
22	*June*	Tongan Islands
23	*July*	New Hebrides
24		*Portsmouth*
25	*October*	Norfolk Island
26	*November*	New Zealand
27	*November*	Sailed for Third Antarctic ice-edge search
28	*December*	Cape Horn

* Italics indicate separate track of 'Adventure'

Voyage – continued

KEY	DATE	POSITION
	1775	
29	*January*	South Georgia
30	*January*	South Sandwich Islands
31	*March–April*	Cape Town
32	*May*	South Atlantic, St. Helena, Ascension Island
33	*July*	Portsmouth

Objects

To search south of the Cape of Good Hope and to sail as close as possible around the South Pole in search of the Southern Continent

To chart and take possession of any land or islands found

Achievements

Proved a continent did not exist in the Southern Oceans

Explored and charted accurately many of the islands in the South Pacific

Illustrations

The possibility of a Great Southern Continent not having been entirely eliminated by his first voyage, Cook agreed to make another to settle the question once and for all.

Young Mr Banks was equally keen and set about collecting a brilliant team of naturalists and artists. Solander, the professional botanist, was again to accompany him with James Lind, astronomer and physician, and two natural history draughtsmen, also Johann Zoffany, the well-known genre- and portrait-painter, and John Cleveley, the marine artist. But the additional accommodation for this retinue made the RESOLUTION unseaworthy and, when it was removed, Banks withdrew his party and took it off to Iceland.

Thereupon, the Admiralty appointed William Hodges as artist to the voyage and, as naturalists, Johann R. Forster and his son Georg, both men of great ability but, as it transpired, difficult shipmates. William Wales, the astronomer and meteorologist, the only one of Banks's nominees to be chosen, quarrelled with them.

In England, society saw many of Hodges's paintings of the Pacific displayed publicly in London after his return, but the world at large learned more of the southern hemisphere from engravings made from his works and the Forsters' for the various published accounts of the voyage.

The Ice Islands, 9 January 1773
Engraving by B. T. Pouncy after a drawing by
William Hodges
*The icebergs of Antarctica provided 'the most expeditious
way of watering . . .' and the crews of the ships' boats are
shown collecting the ice to replenish their water supply*

Table Bay, 1772
Oil painting by William Hodges
This view, as seen from the deck of the RESOLUTION *was
painted in November 1772. The* ADVENTURE *can be seen
anchored farther inshore*

Monuments on Easter Island
Oil painting by William Hodges
Although the monuments were of great interest to the scientists, Cook only spent a short time anchored off the island in March 1774 due to its lack of resources

The RESOLUTION and ADVENTURE at anchor in Matavai Bay, Tahiti
Oil painting by William Hodges
During the second voyage to the Pacific, Cook visited Tahiti twice – in August 1773 and April 1774 – and re-occupied the fort at Point Venus which had been built on the first voyage

The last two pages of Cook's journal of his second voyage of circumnavigation 1772-1775
The last entry gives details of his final landfall and the error of Mr Kendall's watch as being only 7′45″ of longitude

Third Voyage
1776-1780

Ships

HMS RESOLUTION

462 tons

Captain James Cook, FRS
8 officers and 102 men

John Webber, artist

Total 112 men

HMS DISCOVERY

298 tons

Captain Charles Clerke
5 officers and 61 men

William Bayly, astronomer
One servant

Total 69 men

In the Arctic, 1778
from a pen, ink and watercolour drawing by
John Webber
'*The* RESOLUTION *beating through the ice, with the*
DISCOVERY *in the most eminent danger in the distance*'

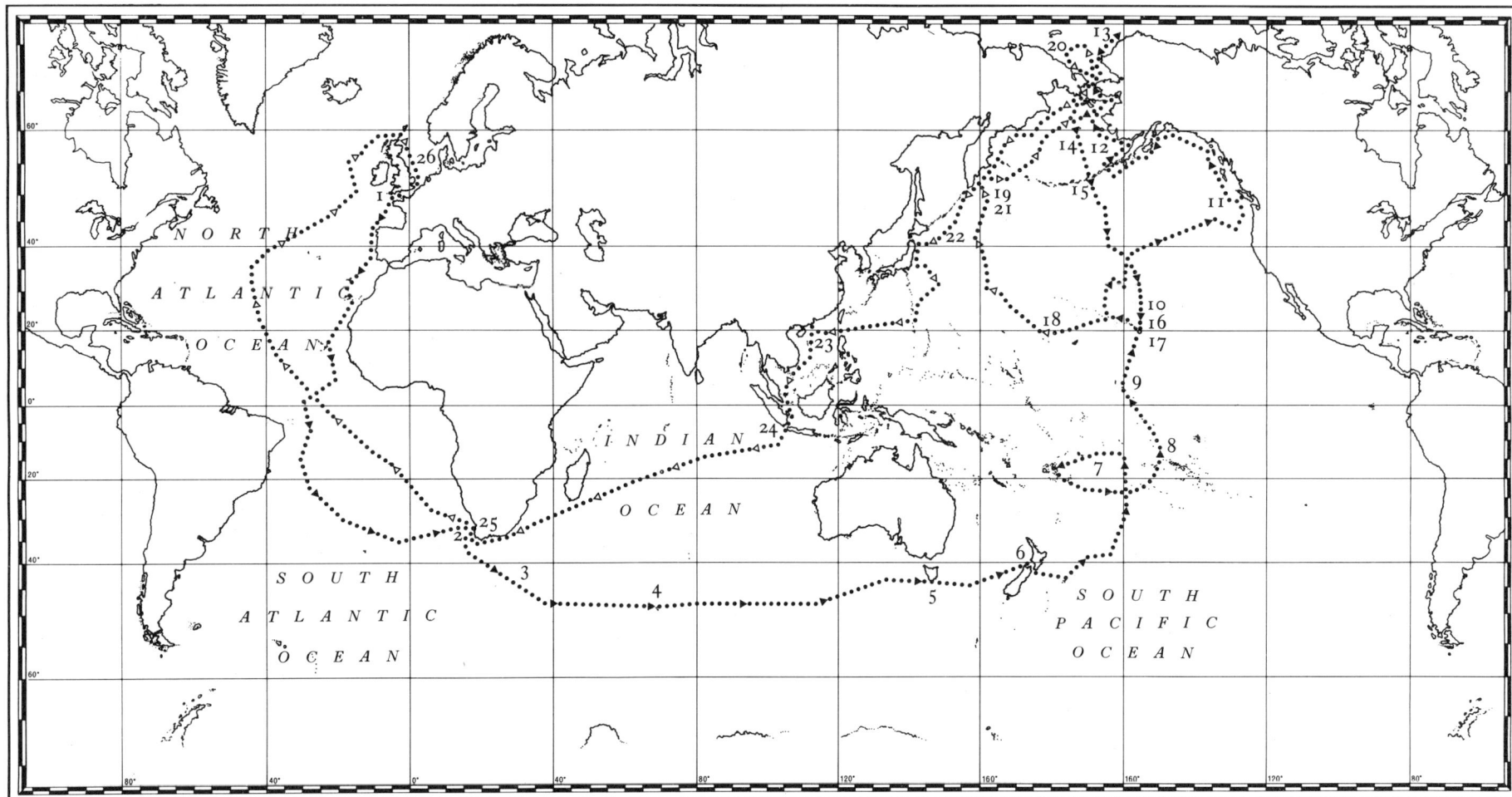

John Webber, artist (about 1752-1798) *left*

Captain James King, RN (1750-1784) *centre*
Engraving by I. Hogg after S. Shelley
He took over command of the DISCOVERY *in August 1779 after Captain Clerke's death*

Lieutenant Henry Roberts, RN (about 1747-1796) *right*
Pastel by unknown artist
He sailed with Cook in the RESOLUTION *on the second and third voyages and drew the general chart showing the tracks of all three voyages, used on the cover of this booklet*

Voyage

KEY	DATE	POSITION
	1776	
1	July	Sailed from Plymouth
2	October	Cape Town
3	December	Verified French discoveries south east of Cape of Good Hope
4	December	Kerguelen Island
	1777	
5	January	Tasmania
6	February	New Zealand, Queen Charlotte's Sound
7	March– August	Explored Central Pacific Islands including Tonga (June)
8	August	Tahiti. Sailed for North American coast
9	December	Christmas Island
	1778	
10	January	Hawaii
11	March	Nootka Sound
12	June–July	Alaskan Coast
13	August	Bering Straits. Search for passage round North America; result negative
14	September	Sailed for Hawaii
15	October	Aleutian Islands
16	November	Hawaiian Islands
	1779	
17	January – February	Hawaii. Cook killed (14th FEB)
18	March	Clerke in command. Search for passage around North Asia
19	May	Avacha Bay, Kamtschatka
20	July	Bering Straits
21	August	Petropavlosk, Avacha Bay. Death of Clerke. Gore in command
22	October	Sailed for Britain
23	December	Macao
24	February	Batavia
25	April– May	Cape Town
26	October	Arrived Thames via Scotland

Objects

STRATEGIC:

To search for a 'North East, or North West passage, from the Pacific Ocean into the Atlantic Ocean'. (This would give Britain naval control of the Pacific)

On the voyage out, to identify islands of possible strategic advantage discovered recently by the French south-east of the Cape of Good Hope

SCIENTIFIC:

To make discoveries on the northern coasts of the Pacific Ocean

To establish a winter base in the Pacific which would make northern exploration practicable in the summer months

Achievements

Searched to the edge of the Arctic ice without finding a passage

Verified the French discoveries south-east of the Cape of Good Hope

Discovered Hawaii and other islands of the Sandwich Group in the North Pacific Ocean which served as bases for Arctic exploration

Explored and charted much of the northern coasts of the Pacific Ocean

Illustrations

In the course of seeking a northern sea passage between the Pacific and the Atlantic Cook was to survey, make charts, and take views of such bays, harbours, and different parts of the coast, and to make such notations thereon, as may be useful either to navigation or commerce; he was also to report on the fauna, flora, fishes and soil on new coasts and 'to describe them as minutely, and to make as accurate drawings of them as you can'.

As a result of the experience of the previous two voyages of non-naval men on board ship, the Admiralty curtailed their numbers further. The Surgeon's Mate on board the RESOLUTION, William Anderson, became the expedition's naturalist, his assistant, William Ellis, the natural history draughtsman.

However, a professional artist was deemed essential and John Webber was chosen 'that we might go out with every help that could serve to make the result of our voyage entertaining to the generality of our readers, as well as instructive to the sailor and scholar.'

Webber was, therefore, to serve, in effect, as an official press photographer would to-day but, instead of taking photographs was to make 'drawings of the most memorable scenes of our transactions . . .'

The voyage was, in fact, the most profusely illustrated of any pre-photographic one. Moreover, Webber had the unique experience of depicting the world between bleak Kerguelen Island in the South Atlantic and ice-packs within the Arctic circle, in all the extremes of geographical variety and human society.

Portable observatory, 1776
A design by William Bayly
Portable observatories were taken on all the voyages for setting up ashore. This is an engraving of the pattern used on the third voyage

A view of Queen Charlotte's Sound, New Zealand, March 1777
Aquatint by John Webber
Cook discovered and named the Sound on his first voyage and subsequently returned there on his second and third voyages. The portable observatories can be seen set up on the shore

Astronomical Regulator Clock, 1769
by John Shelton
Almost certainly one of the actual clocks taken on the second and third voyages. Used ashore with the sextant to determine accurate longitudes and to check the going of the marine timekeepers

Property of the Royal Society, London

A view at Anamooka, Friendly Islands, 1777
Engraving by W. Byrne after John Webber

An astronomical quadrant in use, 1777
A detail from the above engraving
Bayly can be seen observing with his quadrant, which is placed on a cask filled with wet sand

1 ft Astronomical Quadrant, about 1768
by John Bird
Said to be one of the actual quadrants taken on one or more of Cook's voyages. Used ashore for finding accurate latitudes and for checking the going of the regulator clock

Property of the Science Museum, London

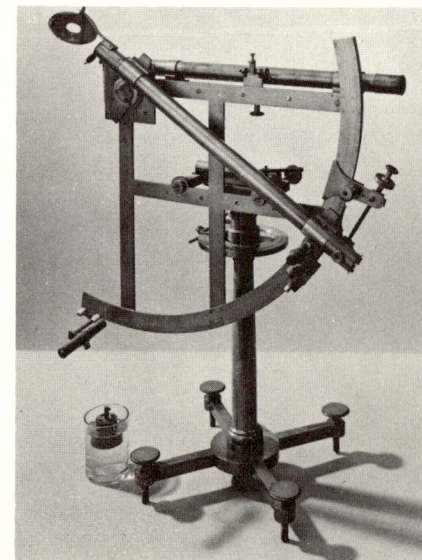

A young woman of the Sandwich Islands (Hawaii)
Engraving by J. K. Sherwin after John Webber

Poedooa, 1777
Oil painting by John Webber
*The nineteen-year-old daughter of Orio, Chief of Ulietea (Raiatea),
in the Society Islands*

The RESOLUTION and ADVENTURE at Nootka Sound, Vancouver Island, 1778
Pen and wash drawing by John Webber
*The ships anchored here for a refit in April 1778 and the forge 'set up to make the
iron work wanting about the foremast' is situated on the beach at the left of the
picture. Observations were also made and the tents and instruments can be seen on
a rock further round the cove*

A white bear
Engraved by Mazell after John Webber

The death of Captain Cook, Hawaii, 14 February 1779
The figures engraved by F. Bartolozzi, the landscape by W. Byrne after a drawing by John Webber
*Cook had gone ashore in the pinnace with a party of Marines to take Chief Terre'oboo hostage
until the* RESOLUTION'S *stolen cutter was returned. The natives became alarmed and in the
general melee that followed Cook and several of the Marines were killed*

Pacific islanders

Settlement

The Pacific islands (with the exception of the Galapagos) were occupied by 'one-way settlement', almost entirely in a west-to-east direction, and by people originating from East Asia. Settlement was not completed in Polynesia until about 1500 AD.

As Cook surmised, this was usually the result of accident – canoes on passage between local islands being blown off course in a storm and, by luck, finishing up at some unknown island, often hundreds of miles away.

Voluntary (or through exile) 'one-way settlement' when done was in big double canoes carrying men, women, plants and animals. When distant islands, such as Hawaii, New Zealand, Easter Island, 1,000 to 1,800 miles from the nearest other inhabited land, were so settled, it was through luck not skill that the voyages ended in settlement and not in disaster.

Navigation

Piloted voyages were limited to islands within a group, or adjacent groups lying east or west, using favourable seasonal easterly or westerly winds; for example, between Tahitian islands and the Tuamota group (180-230 miles).

Pilotage

Pilotage was by dead reckoning, depending for accuracy upon intimate knowledge of local winds, waves, currents and islands, acquired by long apprenticeship at sea within the area where inter-island voyages were made for war, food or barter. The Sun and horizon stars were used also to aid

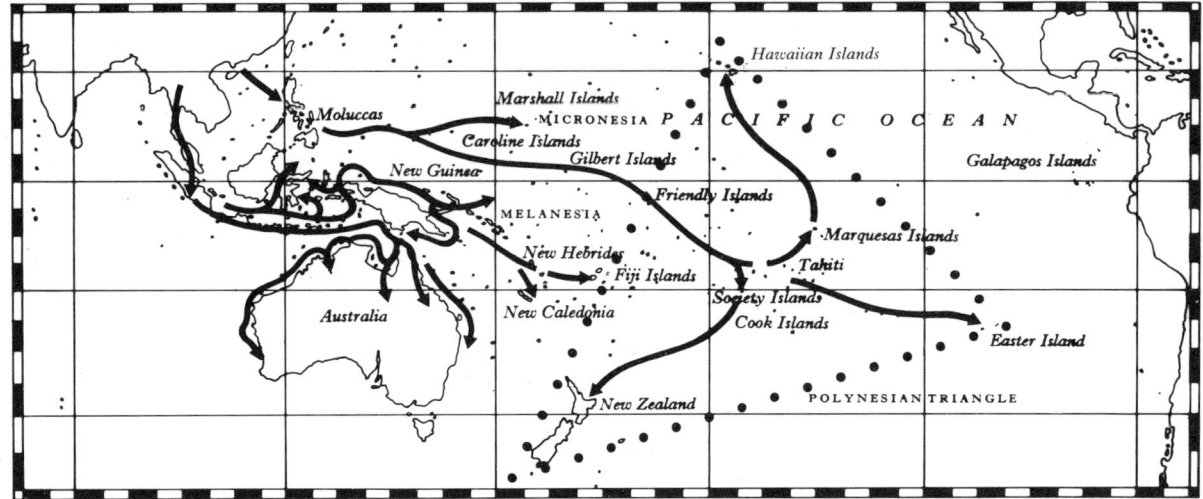

direction keeping. When the sky was obscured, the pilots were *bewildered, frequently miss their intended port and are never heard of more*, as Cook reported.

Culture

The *livestock* and almost all of the *cultivated plants* stem from Asia, not America. *Pottery* was unknown east of the Marianas (Ladrones). *Tools, fishing gear* and *weapons* were of wood, stone and bone, (metal was unknown), and stem ultimately from Eurasiatic sources. *Bark cloth* was made in place of textiles.

The Australian aboriginal culture was most primitive and comparable to the Mesolithic culture of north-western Europe of between 5,000 and 10,000 years ago, that is, food-questing and the use of simple stone chopping and scraping tools.

Tahitian double canoe, described 1769
DRAUGHT, PLAN AND SECTION OF THE BRITANNIA OTAHEITE WAR CANOE
Engraved by W. Palmer

Cook describes the construction of these large war canoes in his journal of the first voyage for 12th July, 1769. '... two Canoes are placed in a parallel direction to each other about three or four feet asunder securing them together by small logs of wood laid a Cross and lashed to each of their gunels, thus one boat supports the other ... some of which will carry a great number of men by means of a platform made of bamboos or other light wood the whole length of the Proes and considerably broader, ... Upon the fore part of all these large double Proes was placed an oblong platform about 10 or 12 feet in length and 6 or 8 in breadth, and supported about 4 feet above the Gunels by stout carved pillors: the use of these platforms as we were told are for the Club men to stand and fight upon in time of battle.'